FOUR AND A HALF DAN[illegible]

Also available from Oxford:

Selected Poems 1956–1986 (OUP, 1987; reissued 1989)
The Other House (OUP, 1990)

Four and a Half Dancing Men

ANNE STEVENSON

Oxford New York
OXFORD UNIVERSITY PRESS
1993

Oxford University Press, Walton Street, Oxford OX2 6DP
Oxford New York Toronto
Delhi Bombay Calcutta Madras Karachi
Kuala Lumpur Singapore Hong Kong Tokyo
Nairobi Dar es Salaam Cape Town
Melbourne Auckland Madrid
and associated companies in
Berlin Ibadan

Oxford is a trade mark of Oxford University Press

First published in Oxford Poets
as an Oxford University Press paperback 1993

British Library Cataloguing in Publication Data
Data available

Library of Congress Cataloging in Publication Data
Stevenson, Anne, 1933 Jan. 3–
Four and a half dancing men / Anne Stevenson.
p. cm. — (Oxford poets)
I. Title. II. Series.
PR6069.T45F68 1993 821'.914—dc20 93-20421
ISBN 0-19-283164-X

1 3 5 7 9 10 8 6 4 2

Typeset by J&L Composition Ltd, Filey, North Yorkshire
Printed in Hong Kong

Acknowledgements

Thanks are due to the periodicals and anthologies in which versions of these poems have appeared: *PN Review, The London Review of Books, The Times Literary Supplement, Poetry Review, Poetry Wales, Stand, The Cambridge Review, Poetry Ireland, Partisan Review, New England Review, The Breadloaf Anthology, 1993, Chelsea Hotel, The Independent, Rialto,* and *The New Yorker*.

'Binoculars in Ardudwy' was first published in *The Poetry Book Society Anthology 1* (ed. Fraser Steel); 'Washing the Clocks' in *The PBS Anthology 3* (ed. William Scammell).

'Four and a Half Dancing Men' appeared in the British Council's *New Writing 2* in 1993.

Among the many friends who have helped to revise and pare down this collection, I wish particularly to thank Matt Simpson, Mairi MacInnes, Mark Elvin and my (ever patient) editor, Jacqueline Simms.

Contents

III

Nature is not intrinsically anything that can offer comfort or solace in human terms—if only because our species is such an insignificant latecomer in a world not constructed for us.

—Stephen Jay Gould

Salter's Gate

For Peter and for Guinness

There, in that lost
    corner of the ordnance survey.
Drive through the vanity—
    two pubs and a garage—of Satley,
then right, cross the A68
    past down-at-heels farms and a quarry,

you can't miss it, a 'T' instead of a 'plus'
    where the road meets a wall.
If it's a usual day
    there'll be freezing wind, and you'll
stumble climbing the stile
    (a ladder, really) as you pull

your hat down and zip up your jacket.
    Out on the moor,
thin air may be strong enough to
    knock you over,
but if you head into it
    downhill, you can shelter

in the wide, cindery trench of an old
    leadmine-to-Consett railway.
You may have to share it
    with a crowd of dirty
supercilious-looking ewes, who will baaa
    and cut jerkily away

after posting you blank stares
    from their foreign eyes.
One winter we came across five
    steaming, icicle-hung cows.
But in summer, when the heather's full of nests,
    you'll hear curlews

following you, raking your memory, maybe,
        with their cries;
or, right under your nose,
        a grouse will whirr up surprised,
like a poet startled by a line
        when it comes to her sideways.

No protection is offered by trees—
        hawthorn the English call May,
a few struggling birches.
        But of wagtails and yellowhammers, plenty,
and peewits who never say *peewit*,
        more a minor *go'way, go'way*.

Who was he, Salter? Why was this his gate?
        A pedlars' way, they carried
salt to meat. The place gives tang to
        survival, its unstoppable view,
a reservoir, ruins of the lead mines, new
        forestry pushing from the right, the curlew.

From My Study

With the hot sun palming my back
like a good therapist,
kneading softly through this
unwashed window, with its
view of stout chimneys
threading smoke in cat-tails,
a maypole of wires joining
pebble-dashed houses, shared roofs
shrunken under frost far whiter
than my neighbour's sagging line
of T-shirts, underwear, sheets
windlessly insisting on their
owners being indoors
cooking dinner, or shambling outdoors
over streets the ice-fur
never quite relinquishes;
old men with old dogs
converging by luck or accident
at corner shops or lamp posts,
huddling, stamping,
huffing on blue nails,
their small warm talk
condensed in vapour trails,

I can't help being again
caught up in it, the marvellous
banality of seeing as it is
the on-going act—just exactly what
slips by our whole attempt to grab it,
press it from living into paying,
the sitcom nobody witnesses, that isn't
funny, except as the real thing
can be funny: Chris in our
alleyway of decrepit fences,
carrying his speakers, mike
and recording equipment carefully
from Pat's house to Jenny's,
and the next day,

moving it all back again from
Jenny's into Pat's—okay,
in the end, even Chris
thought it was funny.

And there's, don't forget it,
the unfunny—call it dead,
grinding nothing-to-do
the lads won't tolerate,
gathering in twos, in sixes and sevens,
to slouch in front of *Spar*,
sharing fags by the memorial's
grey sodden poppies—
you can't say enough in graffiti,
SPAZ, THE CRAZIES, SPUNK, WE HATE PIGS,
the language of hopelessness
has to keep moving.

Arrive with one husband,
skedaddle with another, it won't
go unremarked, but in Langley Park,
where's the accusing finger?
Keeping whispers under wraps—
it's a local folk-art,
the very vicars bicycle in married
and stumble out divorced.
With the century in terminal
spate, a trickle of 'life style'
does finally creep
even here, through the coal dust,
lifting a scum of ash
from the look-alike yards,
cats in, cats out, disputing
each others' flowertubs;
each soiled day arriving,
tea-coloured, sulky,
breaking over, sweeping away
thread by blackened thread
from the enduring village
its clinging web of

history, coal
and its abandonment;
three, four generations
survived by rooms
like this one, two
up, two down, the same
continuous roof
on the teeming furrow,
protective of family television
and microwave convenience cooking;
impregnable air-raid cement
converted into bathrooms,
the netties knocked down now,
rebuilt as garages, dark-rooms,
workshops, potting sheds.

Not, of course, affecting these
fine winter mornings, ice locking
puddles into pleasurable pockets,
the estate running out
along the doggy uprooted railway
where Compass Motorhomes & fumes
have colonized the cokeworks,
ribboning through the playground,
the football field, the allotments—
shacks, goats, cabbages, onions,
pigeons, unsavoury ponies—
out into the green, medicinal
half-planted country;
fountains of hawthorn looking
festive with plastic, unravelled
cassette-tapes, sweet-wrappers,
crisp-wrappers, non-reusable hardware
dumped by refurbishing households,
whose high-pitched runabout kids
gun each other down
from the war's dead bunkers.

Why does that flock of jackdaws
lift in a body?—
a black knotted old-fashioned veil
floating from an aunt's green hat—
now it settles in a dingier field.

Why settle for it? Why here,
with the walking wounded?
Langley Park. Theme park
on the art of failure
for the left-behind,
the old, the stubborn,
the unlucky, the resentful, the gifted
who don't fit into the times, the shifty
poacher who nails up his catch
in his shed, Jim for how long
invalided out, shut up in himself,
in his dignity,
while friendly Alice puckers
with gossip, who used to work
in the meat factory, coming home,
bags full of it, to cook and grin
and marry three peony daughters
on a lifetime's savings.

Choose Langley Park and hunch
in the weather of failure,
wind that blows whiningly
as failure, rain that falls
steadily, unforgivingly, the sort of
failure that lets you off towards evening,
telling you plainly that
somebody famous you used to
dream of becoming
never, of course, will become.
Just the same,
the day can be got through,
filthy air, filthy everything,
the sun splashing special effects
behind the pylons, the *Ram's Head* open,

pumps bowing, Guinness overflowing.
Gary wipes the glasses and plugs in
the organ. The TV streams on.
Linda, in heels, jeans and sleeveless
plunging neckline, shivers at the bus-stop.
Ray donates mixed grills—
chops and blood-puddings for six—
to the charity raffle,
to which James contributes, framed,
four exquisitely engraved
meditations on landscape.

A place to fall back to, thanks,
when you've paid your youth for
visions in the Himalayas. Or when, at 40,
you fall in love with the cello
and give up theology for Bach; or
after that excruciating marriage; or
after that year when you couldn't
for the life of you stop crying,
they wouldn't let you out,
the rejection-slips accumulating,
the girlfriends not even writing.

Thucydides by Virgil by Juvenal—O
Penguin Classics in modern translation,
where else but in Langley Park
does Hermes slide down on a sunbeam
through a thoroughly dirty window,
bearing contemporary bulletins
in bulging satchels?
The Soviet Empire collapses: *the ropes*
are heaved, down come the statues . . .
the fire roars up in the furnace,
the head of the great Sejanus
crackles and melts. Or watch
on inter-temporal television
a funeral production from Attica,
Pericles expounding to the Athenians
the virtues of the Langley Parkians:

We do not feel called upon
to be angry with our neighbour
for doing what he likes—
a sixties slogan that forewarns, no doubt,
of universal cock-up in the nineties.

So the plague has arrived,
the Spartans will take advantage;
the nations are in revolt,
the Macedonians will take over,
not yet affecting Dickon, next door,
in the act of suspending
a small cage of bought-in-bulk peanuts
from a rowan tree already
snowy with chaffinch droppings; now
he's squatting by his frog pond
breaking the ice with a trowel.

No script, no camera, I think
we do communicate. Suppose
the Latin poet was right
about the greatest happiness being
freedom not to fear;
could he have been right, too,
about choosing what's left
of the country,
avoiding Rome and its channels,
never attending interviews
or stuffing on world-horrors,
but at the same time
sympathetically resisting
security system salesmen,
PVC promotion, Jehovah's
Witnesses, satellite dishes,
free gifts, easy payment catalogues?

And if he was wrong, these
dogs, tadpoles, pigeons,
ferrets, talents, temperaments,
love affairs, hate affairs,

exchanges of trout for tomatoes,
home-brew for expertise in
organic leek growing or
1950s combustion engines—
not to say sexy tattoos,
babies, more babies,
books, batik what other local,
threatened consolations—all
have, for a time, composed
the substance of something
people call their lives.

Cold

Snow. No roofs this morning, alps, ominous message
 for the jackdaws prospecting maps of melt.
Something precipitates an avalanche;

a tablecloth slips off noisily
 pouring heavy laundry into detergent,
a basin of virgin textiles, pocked distinctively

with crystals. Your shovel violates this *blanchissage*
 with useful bustle, urgency
pretends, helpless as the swallowed road on which

the air lets fall again a lacier
 organza snow-veil. Winter bridal
the muffled dog fouls briefly. *Don't the cedars*

look beautiful, bent under clouds of fall?
 And it's true, time has no pull
on us; we set it aside for another

'very serious and fundamental' briefing:
 chaffinches at the birdfeed, a sentinel
jackdaw on exposed slates, worried men

tiptoeing their accelerators, deepening
 very carefully each other's ruts.
As if—for how long?—matter had beaten them

and the cold were bowing them back to—
 or forward to—a steadier state. Ice
sets in and verifies the snow.

Imagine a hidden rule, escaped from words,
 stealing the emergency away from us,
starving the animals eventually; first, the birds.

Skills

Like threading a needle by computer, to align
the huge metal-plated tracks of the macadam-spreader
with two frail ramps to the plant-carrier.
Working alone on Sunday overtime,
the driver powers the wheel: forward, reverse, forward
centimetre by centimetre . . . stop!

He leaps from the homely cab like Humphrey Bogart
to check both sides. The digger sits up front
facing backwards at an angle to the flat,
its diplodocus-neck chained to a steel scaffold.
Its head fits neatly in the macadam-spreader's lap.
Satisfying. All of a piece and tightly wrapped.

Before he slams himself, whistling, into his load,
he eyes all six, twelve, eighteen, twenty-four tyres.
Imagine a plane ascending; down on the road,
this clever matchbox-toy that takes apart,
smaller, now smaller still and more compact,
a crawling speck on the unfolding map.

Experimental

Could-be day
lifts keys
from icy broth,
handing them
to the world
conditionally.
Not firm enough to name,
these might-be
river systems
draining a high
topography;
skeletons climbing
into shape and use,

the sun
won't make up its mind.
Weak as the moon,
it hides behind its shine,
raising into vagueness
speculative fountains—
conifer, maple, lime,
layers of beech.
Why let them
matter
or be matter?
Ideas alone
can mean so much.

Dangerous
to be born. Yet
here's a birch
establishing identity
in one chalk-stroke.
Now the black crooked
confluences
of an oak.
A line of pylons
tries on existence,
wiring up
communication.
Does it work?

Brueghel's Snow

Here in the snow:
three hunters with dogs and pikes
trekking over a hill,
into and out of those famous footprints—
famous and still.

What did they catch?
They have little to show
on their bowed backs.
Unlike the delicate skaters below,
these are grim; they look ill.

In the village, it's zero.
Bent shapes in black clouts,
raw faces aglow
in the firelight, burning the wind
for warmth, or their hunger's kill.

What happens next?
In the unpainted picture?
The hunters arrive, pull
off their caked boots, curse the weather
slump down over stoups . . .

Who's painting them now?
What has survived to unbandage
my eyes as I trudge through this snow,
with my dog and stick,
four hundred winters ago?

Negatives

1

Condensed stillness lit meanly
  at four in the morning.
Streetlights a broth of murk,
  amber, unhuman . . .
As I start to describe them,
  horror begins to trust me.

2

A cheap gold-plated wristwatch,
the weakest hand counting seconds
between one quartz minute and another.
One Two Three Four Five.
Spastic stroll without sound.
Count five to put these words on paper.
Or detonate the eclipse of someone.

3

Smoke is my neighbour's crop. His plot
is one of eight clay pots sharing a chimney.
Sixteen chimneys grow smoke along the terrace.

Look. The wind makes a cat of the smoke;
a scuttling black cat, poured velvetly on slate,
filters through the rooks' aerial look-out posts

to where my sad-eyed neighbour stares heavily
at my fence over his fence. He won't watch me
watching him toss crumbs to the clever rooks.

4

Television's virtuous wiles.
  The immortal sheen
of the lovely presenter. She
  wraps up an African massacre.
Now a panel of benign
  communicators . . .

The waxen head on the pillow
  no longer sees. She
keeps the set on for company,
  but can't press
the digits on her other friend
  the telephone,
silent mostly, on the bedside table.

5

What appeared to be another
  filthy rag in the moist alleyway
was underfoot slime, cold
  nucleic porridge, though
one webbed foot protruded
          and came away
     spelling 'frog's leg'.

6

Two whole scrubbed pigs hanging
  hock to hock
from hooks in the ceiling: art
  in a butcher's display,
two evolutionary masterpieces
  minus identity.
Pieces of nature, not yet part
  of the economy.

7

This dream that names itself 'Landsend Beckoning'
  is windless, unwooded, surrounded by ashy beaches.
A raw spit of land severs us from the sea.
  A white-haired black girl invites me to explore.
We have been warned of snakes, but see none—
  only a cave's thready breath and bituminous odour.
Long waves wash sideways over the sand,
  ribbed and intersecting. 'They are like leaves,'
she says. 'Do you remember leaves?
  Now we must learn to swim in their oily rainbows.'

Four and a Half Dancing Men

She knows how to fold
and turn the paper,
guiding the scissors with care
to create for her son
five little dancing dolls.
Toe by toe, hand in hand,
ring a ring a roses,
watch them caper

across the plain and up,
up over the mountain,
five happy men
to amuse a small boy in bed.
So cross. So bored. For
all that, a little blond god,
with the shifting realm
of his risen knees to govern.

The fauna buried in his
landslides, the cities
swallowed by his earthquakes
no longer divert him.
He monitors the marching
of five chained men
with silent intensity,
grave as his liquid eyes.

Up and down, up and down,
his to command,
one, two, three, four
manikins spring by.
He tears from the fifth
an arm, and then a thigh.
The troupe trips on,
though sagging at one end.

Four and a half dancing men.
And the half he made
with an act of his hand
seems to please him best.
He smiles. The same
can be done with the rest.
Four blind men, and a half,
unafraid, unafraid.

Washing the Clocks

Time to go to school, cried
the magnifying lens of the alarm clock.
Time to go home now, the school's
Latin numerals decided.

Days into weeks, months into years.

It's time now thoroughly to wash
the outsides and insides of the clocks.

The broken clocks line up along the dresser,
worn out, submitting patiently.
An old woman in a yellow head-square
prepares to take them apart.

First she pries the glass off a black clock.
The glow-painted arabics fade as she scrubs.

Next, with a little lead key, she
applies herself to a school clock.

Tears must have rusted the hinge.
She has to force the case open.
Two pointed swords and a needle
clatter to the tessellated floor.

Where have they gone? Look for them.
Feel for the hands in the dust,
in the blowing sand. Finger by finger
the numerals break off and drop down.

How competently she's removing
the scarred blank face of my old school clock.

Behind it, the whirring machine,
gleaming brass rods and revolving cogs
making up time by themselves,
rinsing the mesh of their wheels in mysterious oil.

The Professor's Tale

(New England)

It's best, if you can, to love your children,
but adjusting to their strange dispositions
in a nasty world means discipline.

I knew a widow, a PhD, steady and musical,
whose daughter, at ten, suddenly
without warning or reason, stamped on her violin.

I think she had a funny son, too—
made a habit of knifing the piano,
throwing Hi-Fi's and typewriters out the window.

Naturally, with such children, she
came to me for advice. It was not I,
however, who told her to tip them off her life.

She was someone I admired for her agenda.
Not a day slipped by without improving itself
culturally or horticulturally under her eye.

I used to stop by on Sundays—Lapsang, *langoustes*,
a discourse on *Rosa semperflorens*. Beautiful house,
cool and feathery, like her flowers, even in August.

One day, her eyes wet stars, she laughed and confessed,
'All I want is appreciation. When I put myself into a project
I'd like something back in return.' Well, of course.

Practising in life what in theory she knew to be
practical. Maybe subtle. Her garden embodied
espressivo what *pianissimo* she needed to tell me.

'Not that not to do well is wrong.
But that not to be seen to do well
is inexcusable.' I remember her pretty Sèvres,

quaint little antique cups with pink roses.
And those charming photographs of her children
smiling from the grand piano, at appealing ages.

Late

Haunting me at midnight,
a harvester fly, as if
mating with its shadow on the
white ceiling, tells
horror stories about angels.

Its faceless dance has no
individuality. Describe
the identity of that harvester fly?
You can't conceive of such a thing.

Frantic oarlike wings and
six long jointed legs—
awkward to carry—
make a star of it.

For love of light
it immolates its body. No,
it dies in the spider's net—
a root now in the undersoil
of being, not being.

To write of this is the art
of someone who will not
arrive at selflessness,
the fly's, the spider's.

But look at my generosity,
for am I not the one
who provides the ceiling?
And the word for ceiling?

Is there anybody there?
In the fan of stars
opening beyond the ceiling?
In the bloated spider,
in the knot that's now the harvested fly?

Politesse

A memory kissed my mind
  and its courtesy hurt me.
On an ancient immaculate lawn
  in an English county
you declared love, but from *politesse*
  didn't inform me
that the fine hairs shadowing my lip
  were a charge against me.

Your hair was gods' gold, curled,
  and your cricketer's body
tanned—as mine never would tan—
  when we conquered Italy
in an Austin Seven convertible,
  nineteen thirty;
I remember its frangible spokes
  and the way you taught me

to pluck my unsightly moustache
  with a tool you bought me.
I bought us a sapphire, flawed,
  (though you did repay me)
from a thief on the Ponte Vecchio.
  Good breeding made me
share the new tent with Aileen
  while you and Hartley,

in the leaky, unpatchable other,
  were dampened nightly.
If I weren't *virgo intacta,*
  you told me sternly,
you'd take me like a cat in heat
  and never respect me.
That was something I thought about
  constantly, deeply,

in the summer of '54, when I
  fell completely
for a Milanese I only met once
  while tangoing, tipsy,
on an outdoor moon-lit dance-floor.
  I swear you lost me
when he laid light fingers on my lips
  and then, cat-like, kissed me.

Puritan Days

For Sylvia Kantaris

I wonder what, in those days,
when we all drank so much coffee,
  we wanted to prove?
I drank my coffee neat each morning—
  four or five fierce cups—
the way I drank my whisky in the evening.

The whisky was to lift me,
my chance to fall and then be lifted up.
  The coffee?
I wanted to wake (or did I?)
  to see what I was,
or might have been, without the whisky.

The coffee, perhaps, was punishment.
There was all that suffering to do
  after the coffee
and before the whisky. The suffering I thought
  would save me.
Those were puritan days, you know, for art.

Hans Memling's Sibylla Sambetha (1480)

For Peter Forbes and Diana Reich

I had forgotten the weight of her—
the patient gravitas of her
waxed illumination, speaking through
glazed eyes and lips
covertly to us, the future.

Is her energy contained, constrained
by passionate purity?
Or does a veneer of chastity
tame her for necessary marriage?
Her tranquillity is not quite beauty.

Her hands rest, left over right,
on the ledge between trial
and decision. The veil
dropped stiffly from her headpiece
is not the impenetrable veil.

She is someone observed as certain
looking inward, certainly.
Who can tell if her days were happy,
or if vehemence smouldering within
makes visible innocence or venom?

A Quest

Precocious, in the news at nine or ten,
She hero-worshipped creamy Englishmen,
Doe-eyed like Rupert Brooke, with downy chin,
A type that dies in youth and likes to swim.

An Alpha graduate at twenty-one,
She picked them macho and American,
Big Michelin shoulders with a greasy tan,
Sliced pie from hunk to hips like Superman.

She hit the acid road at thirty, when
Her analyst unearthed a psychic yen
To crumple under hairy, beery men.
Her Id was rampant and bohemian,

But still she married Mr IBM
Whose cash came handy when she scuppered him
And went to Cambridge where she lived in sin
With deconstructed Cath'lic Marxism.

At fifty, menopausal, nervous, thin,
She joined a women's group and studied Zen.
Her latest book, *The Happy Lesbian*,
Is recommended reading for gay men.

Level Cambridgeshire

its islands of England
apportioned by drain
and motorway,

doll's-house villages
that have lost their childhood,
roses called 'peace'

and 'blessing' exclusive to
frilly white cottages under
pie-crust thatch.

Can you hear it?
The wind
or traffic. The low-hummed

roar of 36-tonne
saurian lorries and a
soughing avenue of

eighteenth-century limes sound
the same.
In another film

the heroine escapes with
the hero into rural
Cambridgeshire

circa 1666. A field of
barley, feathered; a fen full
of sky-blue

butterfly flax with
undulations like
the ocean's

rolling right up to the
cameraman's pollen-dusted
loafers.

And when Anthea sets up
her easel to catch
a picturesque

angle of the almshouses in
watercolour, she
scrupulously omits

electrical wiring and
TV paraphernalia that,
in strange time,

connect her to
'the brutish, uncivilized tempers'
of these parts,

the cottagers' corpses
stinking, unburied
by the furrows,

Christ's men in retreat
at the Fever House
at Malton,

there 'to tarry in time of
contagious sickness at
Cambridge

and exercise their learning
and studies' until
such time as

God pleased to make
the city safe again for commerce
and superior minds.

II

Visits to the Cemetery of the Long Alive

Hadrian's

Little soul, gentle and fickle,
Guest and friend of my flesh,
Where will you spend your exile,
Cold, annulled, colourless
Who used to give me happiness . . .

A Sepia Garden

Though you won't look at it,
a flat, generous lawn;
two rows of cypresses, ragged
(these days such places can't afford
a full-time gardener); kitchen plot
with stakes for runner beans;
brambles along a brick wall;
and beyond the washing lines,
a wild place with bearded trees
that must have been an orchard.
You can still pick plums there,
and grey apricots, but today,
pulping the windfalls,
an all-purpose handy-man
is driving a tractor-mower.
Preceding us, he makes it easier
to push you in your chair.

He burrs off, cutting a pale corridor
between old-fashioned beds: sage,
potentilla, buddleia attended still
by peacock butterflies; and now in August,
fringed with misty thickets. Ordinary
lavender, but we have to stop.
Even you, hunched like stone,
your eyes slammed shut, feel it
begging, ordering you: re-collect yourself.

Pick a stalk. Crush it in your fingers.
A scent so strong it stings.
Your eyes water, Who is it
lifting . . . from a drawer, is it?
something you ache to touch
and don't dare. A fire opal.
Gold shoes in tissue paper.
A black lace Spanish shawl
scaly with sequins.
A torn cloth doll; its too-blue

porcelain eyes stare at you
as the album, bound in rust,
cracks open wavy pages.
You wrinkle your nose.
A childhood that was yours, and still is,
rushes out like a nursery rhyme,

disintegrates, a handful of
grey-blue ash, lost in the grass

over which we creep,
the present 'me'
listening with head bowed
to the present 'you',
rehearsing complaints,
canvassing chances
(*I won't last the night*)
for painless, everlasting
release from rotting nails
and puffy, useless ankles,
tussles with the nurses,
hours on the commode,
indignities of shaving, balding,
balancing leaden breasts on a bird's chest—
awful little mucks that make your life.

Did you ever worry as a child
about dying and coming back a ghost?
What age would you dress up in?
Would they give you a choice?
The problem would arise in heaven, too,
though you never subscribed to it.

Is it really you?
I thought I'd never see you again.

You raise it every time, that question of
'really being', a Dorian refrain
or echo trembling in your sepia garden,
drifting through the corridors of tableaux.

Oxford Square: *When I was married*
I'd never boiled a kettle or run a bath . . .
dissolving to thirties' Cambridge and *those*
terrifying academic dinner parties . . .
I, the youngest, least educated, wife
having to lead the ladies out
so the men could sip their port.
Cricket at Fenners: *Fearful swindle,*
Bradman out, bowled first ball . . .
Snow forever talking and making me talk,
putting me verbatim in his book . . .

Saved by the war: *No servants—the relief!*
Our own potatoes, endless rhubarb . . . a fox
got the hens, poor Sue's religious phase, thank God,
hit the rocks, but the nightmare of German invasion,
what would we do? Peter on the Nazi blacklist.
I, a Jew. Could we save the children?
D said, better to kill them. Yes, I said, how?
Carbon Monoxide. Never, we won't have a car.
I can still hear him bellowing,
Don't be so bloody helpless, woman, steal a lorry!

Frayed, recurring clips; can they be you?
Is there a thread of you-ness—not a soul—
interred in those deep sleepy layers?

Mr C's choleric moods, imposed-on sons,
threats, escapes, reprisals,
withdrawn funds. *I always had a*
passion for . . . cream? shells? hill-country
stitched together with stone walls?
Today it's a pre-war saga of how
Miss Adams built her cottage in revenge
when Miss Ellison took up with the
Honourable Griselda Bingham-Wilkes in Wales.
Then minor grievances, medical details,
Dr So-and-so and Matron Such-and-such.
An intolerable way to live!
His hurtful remark; that festering quarrel:
it must be written down, never to forgive.

I can't help supposing that
it never really fitted, your long life—
those dreary, loose-cut, madam
hand-me-down roles: 'rich man's daughter'
followed by 'don's frightened wife'.
Far too baggy for you,
who would have made a devout botanist;
a happy stone mason, possibly;
or an artist . . .

In America, we assume there's a self
like a spine in us. Whatever we achieve,
we construct, let's say out of selfishness.

But imagine someone lecturing *you* on
Establishing Identity through the
Acculturation Process. *Absolute rubbish!*
I can hear the tuba in your voice
dismissing *all that rot about identity*
to the lowest ditch of the ridiculous.

So for you, there's been daily irritation,
the cramped frustration of attempting
the jigsaw with pieces missing. My dear,
I promised to love you, and I do.
But no good woman ever lived harder
to stitch a spirit to! As true:
no one could possess more—surely, it's
still called—'character'—than you.

The wind's up. I see the handy-man
has put away his mower. Time to take you in.
So kind of you to come. Now all the fuss
of blundering to the lift, those awkward doors,
corridor smells, efficient-looking nurses.
Under my cheery words and hasty kiss,
the pressure of your uncathartic tears.
I avert my eyes from other humps in chairs.

Something has to feel, connect, suffer.
Something must be . . .

Think of it as new-mown grass, or picked lavender,
about as much 'self' as they'll save of us
in a gauze pouch, or a rubbed leather book
between pages of mottled photographs
when they throw us away with our clothes.

Bloody Bloody

Who I am? You tell me
first who you are,
that's manners. And don't shout.
I can hear perfectly well.

Oh, a psychologist.
So you think I'm mad.

Ah, just unhappy.

You must be stupid if you
think it's mad to be unhappy.
Is that what they teach you
at university these days?

I'm sure you're bloody clever.

Bloody? A useful word.
What would *you* say, jolly?

It's bloody bloody,
I assure you,
having to sit up
for a psychiatrist—
sorry, *behavioural psychologist*,
I know there's a difference—
when I want to
lie down and sleep.

The only sensible thing,
at my age, is to be
as you well know
dead, but since they
can't or won't manage
anything like that here,
I consider my right to sleep
to be bloody sacred.

I can't hear you,
I'm closing my eyes.

Please don't open the curtains.

I said *keep the curtains shut*!
Thank you.
Hate you?
Of course I hate you,
but I can't, in honesty,
say I blame you.
You have to do your job.

There.
That's my telephone.
How fortunate.
You'll avail yourself
of this opportunity, won't you,
to slip tactfully away.

Hello? Yes,
two pieces of good news.
One,
you've just interrupted a most
unnecessary visit,
a young psychological person
is seeing herself out.
Two,
you'll be relieved
to hear I'm worse, much worse.

Black Hole

I have grown small
inside my house of words,
empty and hard;
pebble rattling in a shell.

People around me, people.
Maybe I know them.
All so young
and cloudy, not . . . not real.

I can't help being the hole
I've fallen into.
Wish I could tell you
how I feel.

Heavy as mud, bowels
sucking at my head.
I'm being digested.
Remember those moles,

lawn full of them in April,
piles of earth they threw
out of their tunnels. Me, too.
Me, too. That's how I'll

be remembered. Piles
of words, sure, to show
where I was. But nothing true
about me left, child.

A Tricksy June

I'm an old woman who wants to die.
    They make me live.
I know a woman who wants to live.
    She's going to die.
Why? Why? It's tricksy, June.
Grass looks greener for never sighting the sun.
    Nights shrink, days stretch.
The nurses have to smile. The dinners come.
    Is a mind
turning down on us or up on us its slate thumb?
    Who do you think orders
these ugly salvia-trimmed gardens to bloom?
    Up north, my home,
pear-blossom's snow on the grass. Must be
    the martin's back.
I'm missing those old-fashioned yellow roses
    hugging my gate,
already losing their frowsy, blowsy furbelows.
    Who lets them yawn
and disintegrate, petal by lazy petal
    without complaint?

Lost

Stone-age, stone-grey eyes
clear in her glove-like skin;
a look of having been ironed
before she shuffled in.

She's cradling a pink blonde doll
in a quilted bag, pink satin.
She lifts it out a while.
She puts it back again.

Her dead child? Poor, poor lady!
We burn to know . . . what reason?
No sign, from mouth or body.
She stuns our pity even.

Mrs Meredith

'The genome is the sum of the genetic information passed down from one generation to the next, a message three billion letters long . . .' —The Economist

As the spirit rejects its bacterial origin,
hah! so do I you, thin female body,
wrung dry, pegged down, shaky
maker of sick protein,
splayed for the surgeon's inspection.

The glare of diagnosis
gets to be a bore.
I can't see why? . . . I wouldn't want
to keep you any more
if I could be sure of pulling out,

cutting loose from your
wrinkled, disintegrating anchor
rising, say, on a puff of cumulus
to some tinkling strato-cirrus
Bliss Forever and Ever Unlimited.

Hah! Dust into dust no longer,
but genome by genome, the dance
of information, dear body,
links matter eternally to matter.
In the chains of continuance

where will my self be?
A traffic of deafening cells
pounds through me. And those leafy
summer by-ways I used to treasure,
Tweedsdale and its hills?

Ploughed under, with the names
my mother gave me—Mary Louise.
Her name was Florence,
daughter of the Renaissance
with all that music in her

they said I inherited . . . it must
have been you . . . the music
must have come through you, poor body,
to the waste that now is . . . Jennifer,
my daughter, visits me

on Sunday, or whatever day. She's
always busy. A good word, busy.
Music kept me busy, to protect me.
I might have thought too much.
Busyness, better than loneliness . . .

Am I afraid? Not of dying. Not,
science knows, of being without my self.
Afraid, I think, of being *in* life
forever . . . not knowing, without me,
who I am. My tag says I'm Mrs Meredith.

To Witness Pain is a Different Form of Pain

The worm in the spine,
the word on the tongue—
        not the same.
We speak of 'pain'.
The sufferer won't suffer it
        to be tamed.

There's a shyness, no,
        a privacy,
a pride in us. Don't divide us
        into best and lesser.

Some of us, 'brave'? 'clever'?
watch at the mouth of it.
        A woman vanishes,
eyes full of it, into it,
the grey cave of pain;
        an animal drills
unspeakable growth for cover.

Outside, we pace in guilt.
Ah, 'guilt', another name.

        Not to reproach
is tact she learns to suffer.
And not to relax her speechless
        grip on power.

With a Gift of Welsh Poppy Seeds

These hundreds and thousands of golden possibilities
are not, please, to be shut in the pepper grinder,
but scattered on upturned earth with other weeds.

They come with the compliments of fickle weather,
soil that likes saxifrage and winter heather.
Think of us when you plant these seeds.

May Bluebells, Coed Aber Artro

No Greek self-pitying *hetairos*
in blue-rinse curls,
the north's true
Hyacinthus non-scriptus
(much written about, nonetheless)
beloved of Hopkins, who in Hodder's Wood
caught perfectly its
'level shire of colour'
while his companions talked;

West Country 'Crowfoot' or 'Grammar Graegles',
in Welsh translated 'Cuckoo's boot',
'Blue of the Wood', 'Welcome Summer',
each silky delicate bell-stalk
carrying its carillon to one side;
dusky wine-cups, ringers
of creamy anthers; in *Cymbeline* misnamed
the 'azur'd harebell' by Arviragus,

and even in our time,
self-assigned to resurrection.
Camping gas set burning
at the lowest visible flame.
Ice-age giant still nourishing
the trodden mulch and green enchantment
of his daughter beechwood, watering
one more summer out of hazy veins.

Binoculars in Ardudwy

A lean season, March, for ewes
who all winter camped on the hills.
They're gathered in now to give birth
to children more cheerful than themselves.

There's a farmer, Land Rover, black dog
trotting, now rolling on its back.
At the gate, sheep bunched—one alone
drifting down the steep Cambrian track.

Look now, the sun's reached out,
painting turf over ice-smoothed stone.
A green much younger than that
praises *Twllnant* and *Pen-isa'r-cwm*.

All this through the lens of a noose
I hold to my focusing eyes,
hauling hill, yard, barn, man, house
and a line of blown washing across

a mile of diluvian marsh.
I see every reed, rust-copper,
and a fattened S-bend of the river.
Then, just as I frame it, the farm

wraps its windows in lichenous weather
and buries itself in its tongue.
Not my eye but my language is wrong.
And the cloud is between us forever.

Under cover of mist and myth
the pieced fields listen, and whisper
—Find invisible *Maesygarnedd* . . .
Y Llethr . . . Foel Ddu . . . Foel Wen.

The italicised Welsh words are farms in Cwm Nantcol. Y Llethr, Foel Ddu and Foel Wen are mountains.

Trinity at Low Tide

Sole to sole with your reflection
  on the glassy beach,
your shadow gliding beside you,
  you stride in triplicate across the sand.
Waves, withdrawn to limits on their leash,
  are distant, repetitious whisperings,
while doubling you, the rippling tideland
          deepens you.

Under you, transparent yet exact,
  your downward ghost keeps pace—
pure image, cleansed of human overtones:
  a travelling sun, your face;
your breast, a field of sparkling shells and stones.
  All blame is packed into that black, featureless
third trick of light that copies you
          and cancels you.

Painting It In

Remembering Lesley Parry

Wake up at six o'clock. We're out to sea.
Nothing beyond that fence and slatted gate
but a grey wave and plume-like shapes that could be
flaws in the canvas or unmixed pigment in paint.

Stones, blurred poppies, a wheelbarrow full of grass
affirm a foreground. The world must exist out there.
People must be getting up and getting washed,
putting the kettle on, picking up a newspaper.

Somewhere it must matter terribly not to be late,
not to miss the limousine to the airport,
not to be missed when the finance committee votes,
when the training course commences, not be left out.

But somewhere is hard to believe when it's not
invented,
when the world blindly refuses to admit detail.
All that's required is pastoral: sheep among stunted
rowans; for background, eroded 'Moelfre' or 'bald hill'.

The thing's been done so many times. Imagine
brushing the lichen's pale quartz over the rocks,
now the shocking pink foxgloves, painting them in,
old fashioned *belles de joie*, drunk on their stalks.

What if today decides not to take off its veil,
never to palliate art with a grand show
of perspectives up the valley? More likely all we'll
get is light's first lesson, an application of gesso,

a whiteout of air—sweet, soft, indestructible,
the cloud of unknowing reluctant to create the known.
Hills, stones, sheep, trees are, as yet, impossible.
And when things are unmade, being also feels less
alone.

Original Sin

When the cat caught two
nesting swallows
and laid them out for her,
she was minded to kill;
the cat knew it
before she did.
Where is that murderer?

The murderer disappeared.
She? Vanished and became
a tidal wave, white-nailed,
uprising in a claw.
Where was that cat
she would kill
for killing? It all

happened before lunch,
the roar, the wreck
of an old alliance,
the airing cupboard treaty,
peace, and the tidewrack
laying out for memory
the repeatable dead.

Before she threw them to the crows,
she stroked the swallows'
curled feet and ferrous heads.
Wings, folded and oval,
framed them like cowls.

Terrorist

One morning I despaired of writing more,
        never any more,
when a swallow swooped in, around and out
        the open door,
then in again and batlike to the window,
        against which
beating himself, a suicide in jail,
        he now and then collapsed into
his midnight iridescent combat suit,
        beautiful white markings on the tail.

Inside his balaclava, all he knew
        was something light and airy he had come from
flattened into something hard and blue.
        Thank God for all those drafts I used to
scoop, shove or shovel him to the transom,
        open just enough to let him through.

Off he flew, writing his easy looped
        imaginary line.
No sign of his adventure left behind
        but my surprise
and his—not fright, though he had
        frightened me, those two
bright high-tech bullets called his eyes.
        What they said was
'Fight and fight and fight. No compromise.

When the Camel is Dust it Goes Through the Needle's Eye

This hot summer wind
is tiring my mother.
It tires her to watch it
buffeting the poppies.
Down they bow
in their fluttering kimonos,
a suppressed populace,
an unpredictable dictator.

The silver-haired reeds
are also supplicants.
Stripped of its petals,
clematis looks grey
on the wall. My mother,
who never came here,
suggests it's too hot
to cook supper.

Her tiredness gets everywhere
like blown topsoil,
teasing my eyes and tongue,
wrinkling my skin.
Summer after summer, silt
becomes landfill between us,
level and walkable,
level, eventually, and simple.

North Wind Light

I wish you could watch it gilding
the hennas of the marsh. Our summer's over.
Between the iron gate's upright and top rung,
an orb web, jewelled. Ewes have laid more
glistening pebbles, the pasture's
pointillist with dung,
and burnished dung-flies busily feeding.

Teardrops hanging   one, one, one,
from crosswires of twisted fencing.
After heavy rain, a flood of sun.
Now a speckled young partridge—pecking,
looking, pecking.
Fly away, silly bird, fly! Or the pasture
will be grazing on your light bones soon.

TWO POEMS FOR JOHN COLE (1910–91)
AND ONE FOR ANNABEL COLE

Dinghy

Though you won't feel again
    the jib sheet
scalding your hand, or coil
    it again
around the cleat,
    or be able to read
from the shiver
    of a sail
the precise crisis—
    timing the rudder,
coming about,
    reaching at high tilt,
hauled tight
    against the slapping river—
your discontinuedness
    is not yet felt.

All around you
    the life you built
is tacking habitually
    in skilful directions,
swimming your sea,
    climbing your mountains.
But you're left out,
    and grief fills
the leaf-shape of a
    toy boat
far on the horizon;
    while you still weigh
upon these Welsh hills
    alight with your look
of patient inquiry—
    more accurately, dear John,
the attentive
    courtesy of that look.

Cambrian

Here is one more fiery
        sunset for you
not to share
        with the ravens who
rebuild every winter
        their bulky stack
on the ice-cut rock.

Without you,
        the nest will be there
as *Gareg Lwyd*
        will be there—
two eyries folded
        in the lap
of what nineteenth-century
        geologists
used to call *grauwacke*.

Of Murchinson's quarrel
        with Sedgwick
and how they changed that,
        disputing Silurian layers
on the Cambrian map,
        we might have talked
by your Christmas fire,
        we two and you,
while outside, unprovoked,
        the wind blazed higher,
renewing its quarrel
        with everything they drew.

After you left

that perpetual summer sun became unpleasant.
It had an ozone minus feel about it,
everyone short tempered and too hot.

So when the impacted storm gathered and broke,
we ran to the East window to look out.
The cwm had already spumed its rolling ghost,

Moelfre was muffled, the Rhinogs hidden or eaten,
thunder thundered, lightning fired and missed
us—maiming, I think, the electricity grid.

Then welcome rain, provisional, unfriendly.
The sheep stood backward to it, stony still,
looking, or trying to look, invisible.

Nothing electric at *Pwllymarch* presently works.
I'm writing by candlelight. Supper's half cooked
on the fire; the wine is good.

'Happiness shouldn't be so important.'
About four square inches on the canvas
is a piece of luck.

OXFORD POETS